A PRIMER ON RECYCLING

The Philosophy, History & Future of Our Waste

Greg Dudish

Foreword

Something stuck with me after when I went on a tour of the San Mateo County recycling center in California. Everything I discard can be used again! But, a more interesting question is why something I discard cannot be used again. Many hours of research turned into many more hours as the problem of what to do with our waste spanned many fields: chemistry, business, phycology, government, history, etc. Waste became much more complex than I expected. This book is an attempt to bring the complexity into focus and give others a starting point for their own research.

The book is structured to first give a high level view to answer "What is waste?" then how early recycling of that waste evolved into modern techniques with a projection to the future. As a case of a modern recycling facility, I attempt to give you, the reader, a tour of the machines and overall process used at the facility I toured. As with the actual tour, there are a few tips on how to improve how you recycle your waste (Remember: Reduce, Reuse, Recycle!). And lastly, the skeptic's arguments on recycling are address to give an idea of the entire balanced picture along with the challenges faced by the recycling industry.

#1

An Introduction to Waste

How Waste is Created

Recycling is analogous to a manufacturing processes - generally, each process starts with inputs (raw materials) and generates outputs (products & waste). But I didn't realize how we generate waste could also be viewed as sort of a manufacturing process. The waste process starts when someone buys new product (input). The product is used and its value diminishes (unlike manufacturing where value is added during the process) and ends at the final step in consuming a product - garbage/waste (output).

The basic steps are: New Products become Used Products and eventually lose value to become Waste Products.

Generally, the waste production process in a community follows the flow on the next page:

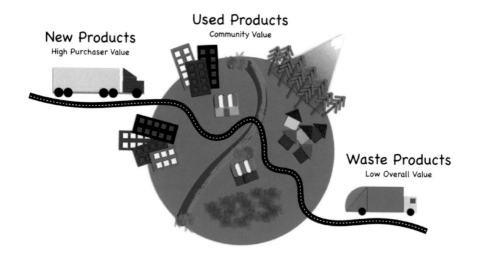

New Products
High Purchaser Value

Used Products
Community Value

Waste Products
Low Overall Value

NEW PRODUCTS come into the community and are sold by retailers. Someone finds value in this item so they buy it to use that value. The new iPhone or a cold Coca-Cola are valuable to the purchaser. After the value is taken by the purchaser, these products become used.

USED PRODUCTS have less value to the initial purchaser. But, an item can still be reused by the purchaser or valued by others in the community. An old iPhone can still be sold on Craigslist to someone else in the community. But, a used can of Coca-Cola would even be hard to give away.

WASTE, ideally, waste occurs when a product is unusable or no longer wanted by the community. Realistically, waste is made regardless of if someone else could still use a product. This waste is collected and hauled away by garbage collectors.

But the value of a product doesn't stop when it is removed from the community!

Resource Recovery - Value in Waste

Resource recovery is all about seeing value in waste. The fundamental challenge is "to return resources to their best and highest use." To understand the value in our waste, let's look at a profile of our waste. Here is a breakdown of typical American waste and the costs/value of that waste recovered by resource recovery companies:

Makeup and Value of Average American Waste

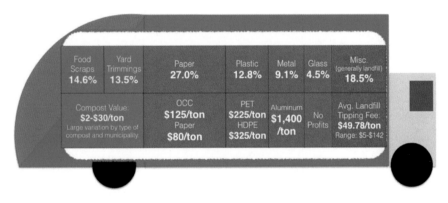

Food Scraps 14.6%	Yard Trimmings 13.5%	Paper 27.0%	Plastic 12.8%	Metal 9.1%	Glass 4.5%	Misc. (generally landfill) 18.5%
Compost Value: $2-$30/ton Large variation by type of compost and municipality.		OCC $125/ton Paper $80/ton	PET $225/ton HDPE $325/ton	Aluminum $1,400 /ton	No Profits	Avg. Landfill Tipping Fee: $49.78/ton Range: $5-$142

ORGANICS (GREEN): One of the earliest collection of compost was yard trimmings. Separating trimmings from trash remains as one of the highest diversion percentages for all waste produced. In 2013, over 60% of yard trimmings produced were diverted from landfill.

Food waste compost had slower adoption in the US. In 1990, only 9 facilities were in operation versus the 1,400 facilities for composting yard trimmings. The compost industry is still relatively new and will develop over time. As of now, resource recovery companies make a marginal amount of money on food waste compost. The value for these companies comes from the diversion of the weight of compost out of the waste stream, therefore the company avoids the fees charged by landfills to dump materials there.

RECYCLABLES (BLUE): Materials which can be recycled hold the highest value because their potential to be remanufactured into new products.

Paper makes up the largest percentage of what enters a recycling center. To increase the value of the paper, it is sorted into various sub-products (cardboard, newspaper, and mixed paper).

Plastics make up the second largest portion of what is recycled. The challenge in unlocking the value of plastics is in sorting – just like paper. There are several different types of plastic, each with different resale rates and end (re)uses. If these different types are sorted, the value becomes greater.

Aluminum revenues boost the entire recycling facility. It has a very small volume but is the most profitable. By weight, it is less than 2% of the recycle stream but it generates 40% of the revenue of the recycling center at a whopping $1,400/ton!

Glass is not usually profitable and depends on the region for resale markets or government subsidies. Its weight makes transportation costly therefore reuse is limited by region. Some states, like California, subsidize the losses incurred with processing glass by charging distributors and consumers a fee per bottle sold. Also, new glass bottles are significantly cheaper for manufacturers than washing and reuse of bottles.

LANDFILL (RED): Miscellaneous contains rubber, leather, textiles, plastic films (the majority of my personal trash), and other items which could not be recycled or easily classified. These materials go into the landfill. And whenever a material is sent to a landfill there is a fee per ton. So, these items actually cost "resource recovery" companies money. These companies feel the pressure of high fees and are incentivized to divert more tonnage from the landfill. The cost of adding to a landfill (known as the "tipping fee") varies widely across the country - the highest is in Washington ($142/ton) and the lowest is in Texas ($5/ton). So it's no surprise the Pacific Northwest has been considered the leader in recycling for the past few decades.

Resource recovery means to focus on each material in a community's "waste" and understand with processing, the material can be separated and reused in new products. Ideally, the same product can be made from the recycled product but most of the time this isn't true. Most recycled materials are "downcycled" into a less valuable product – for example bottles and containers are melted together into a lower grade plastic which can be made into a park bench.

Modern Recycling - aka. Landfill Avoidance

Landfill "tipping fees" (the cost to dump garbage at a landfill) are on the rise and municipalities face pressure to cut these costs. "Whenever there aren't huge open spaces, waste disposal costs are the fastest rising item on any municipal budget," says John Schall, the recycling director at the Division of Solid Waste in Massachusetts. "In the Northeast it's starting to cost so much for disposal that you don't even need to bring in revenues for the recyclables, you just need to get them out of the waste stream." says John Purves.

The answer for municipalities comes in advances in sorting technology. Technology allows for single-stream recycling - where residences can put all recyclables into one curbside bin for pickup. The previous method required residences to separate paper, metals, and plastics at home. Separation before pickup means materials have less contamination and can be sold at a higher price. But, the discipline required to separate materials reduces overall recycle rates. Single-stream simplifies recycling. Using the South Bayside Waste Management Authority (SBWMA) in San Mateo County as a case study, the switch to single-stream increased recycling amounts by 30%. Even with a more contaminated recyclate (reducing its resale value), the overall amount of recycle increased total revenue and profits increased.

The latest "trend" in waste collection is compost. In the Bay Area, compost is now collected in a bin which people put out on the curb. In the past, curbside compost was yard scraps only - now, technology advances allowed an increase in what can be processed. Right now, SBWMA can process food scraps, food soiled paper, and, of course, yard trimmings with the longer term goal to allow people to compost "anything that was a plant."

Future

Many municipalities and "resource recovery" companies have the goal of zero waste coupled with the financial pressure of increased dumping fees or local landfill closures. These will encourage further experimentation in the waste industry and drive technology advances - leading to changes in collection techniques.

One technology making headway in the United States is the OREX Organic Press - or the garbage juicer. Using a hydraulic press, a piston squeezes the moisture (aka. juice) out of the landfill bound stream - thereby reducing the weight of the material sent to the landfill and reducing the cost in tipping fees. The juice consists of organics which is collected and can be sent to a biogas plant to be processed into biofuels or used as compost.

Projection of the future likely will follow the major problem of recycling in the past - as products become more complex and use more materials, those materials become more difficult to separate at end of life and therefore harder to recycle. Both recycling and manufacturing technology will have to advance in order to separate a more complex recyclate and manufacture a wider range of recycled materials.

A notable example of complex recyclate is electronics. Computers are very complex and require several different materials, each with a very high value but only when separated and purified. A whole new segment of resource recovery had to evolve in order to adapt to computers: E-waste.

The Holy Grail

Ultimately, the Holy Grail in resource recovery is separation technology will advance far enough to dig-up existing landfills and extract all the valuable materials which have accumulated over the years. With the materials in the landfills and achieving 100% recycling rate, we theoretically should slowly be able to stop mining and pumping raw materials from the earth to create a closed system on materials, rather than a linear system from the earth, to our houses as products and ending in the landfill. But until the future arrives, we have to become more conscious about the waste cycle and understand what we can do to help.

#3

Case Study:
Shoreway Environmental
San Mateo County, California

When I toured the SBWMA Shoreway facility, my main amazement was machines – not humans do 99% of the sorting for recycling. Not only was I blown away that equipment was able to automatically sort, but it was much "lower tech" than I thought - a lot fewer lasers than I imagined! In an attempt to capture the tour experience, I'll step through each piece of the recycling equipment in the process.

The tour was limited to the recycling area - the other areas of the Shoreway complex were transfer stations for landfill and compost. Collection trucks unload at these transfer stations and the trash or compost is loaded into a tractor trailer which will haul it away to a landfill or composting facility.

Recycling at Shoreway

The first step happens before the truck even picks up recycling. The people of San Mateo County presort into 3 different bins: landfill, compost, and recycle.

After the recycle is picked up, trucks unload onto the floor at the Shoreway facility. The overall process flow is shown below – each of the technologies will be discussed in detail.

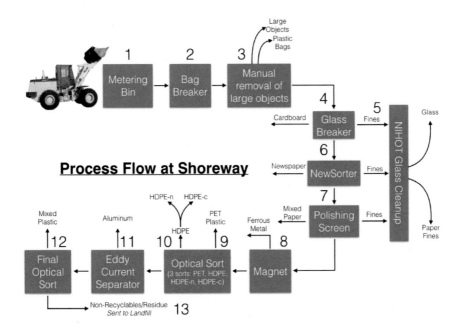

Process Flow at Shoreway

1. Front-end Loader dump material into a metering bin. The metering bin creates a consistent flow of material downstream. A consistent feed ensures predictable conditions for the machinery and provide high quality separations. This is important because if a front-

end loader put material right into the conveyer belt, the material would enter the equipment in "waves" meaning a large amount of material would enter the equipment followed by almost no material while the Loader was getting another load.

2. The material enters a bag breaker which tears apart any plastic bags containing recyclables. This releases the recyclables within the bag.

3. Before material enters the first separation machine, a person removes large materials such as metals or wood pieces that could damage the machinery. The plastic bags from the previous step are also removed since they are not recyclable and can cause problems within the equipment.

4. Material enters Glass Breaker DRS, which shatters glass bottles, dropping the broken pieces through the equipment down with other <2" fines. All other mixed recyclables don't fall through the screens and are not broken by the discs which selectively crush glass. Old Corrugated Cardboard (OCC) coasts over the top and is also sorted out by this machine. The OCC is inspected manually to ensure quality and remove any contaminates from the end product.

5. The <2" fines enter a conveyer belt and are further separated by the NIHOT Glass Cleanup System which separates glass and heavies from <2" lights (paper and other fibers).

6. The remaining stream produced by the Glass Breaker is fibers (e.g., newspaper and mixed paper) and containers (i.e., 3D objects - cans and bottles). These enter the NewSorter which separates ONP (Old News Paper) from mixed paper and containers. The ONP floats on top of the spinning discs while containers and mixed paper fall through. The ONP is inspected manually to ensure quality and remove any contaminates from the end product.

7. The mixed paper and container stream enters the Polishing Screen, which separates the containers from fiber. Fiber floats on top of the screen and over the top while 3D containers float downward, creating the separation. Fines are also produced. The mixed paper stream is also manually inspected to ensure quality and remove unwanted materials.

8. Ideally, at this point in the process, metal cans (such as steel and aluminum) and plastics are left to be sorted. But since no process is perfect there also exists a small amount of contamination of items, which were not properly separated out, or items which should

be sent to the trash. The first step is to separate out ferrous metals (steel) and non-ferrous metals (aluminum) from the stream. First, the ferrous metals (which are attracted to a magnet) are removed. All containers pass under a magnet to remove the ferrous materials.

9. Next PET is removed from the containers stream. There are 2 different ways offered by the company used in designing the Shoreway facility. Reflective and transmissive technology. Reflective, puts light onto the bottle and measures what is reflected back to the detector overhead. Whereas transmissive measures what light is passed through the container. Each type of plastic has its own signature from these techniques and the machine then puffs air to sort the materials from one another.

10. Another sorting machine removes HDPE plastics from the containers stream. Removed HDPE is sorted again by another machine into HDPE-n (natural – example a milk jug) and HDPE-c (colored – Tide detergent bottles).

11. Next the aluminum is removed using an eddy current technology. Since aluminum is a metal but non-ferrous, normal magnets will not work. It's like magic to watch the aluminum fly out of the recycling stream!

12.The remainder is a mix of plastics and non-recyclables. A final separation occurs to remove plastic materials from the non-recyclables.

13.The non-recyclables are sent to the landfill.

After all the separations, each material is shipped to manufacturers to process as raw materials into new products.

#4

Seven Things You Can Do

#1 – Mindset:
"I will act as if what I do makes a difference." - William James - painted on the wall at the entrance to the Shoreway Educational Center

Shift your mindset about your personal waste generation and realize what you do matters.

#2 - Reduce:
Be mindful of your purchases. Waste production is a linear process, whatever is purchased is eventually discarded. So, reducing the amount consumed per person will have the largest impact. Between the 1960s and now, Americans have almost doubled the amount of waste produced per person!

Waste Per American *Doubles*

1960   Now

#3 - Reduce Food Waste:
Practice mindful purchasing with food. If your eyes are too big for your stomach at the grocery store and the food molds that is food waste and it makes up a large percentage of our garbage. So large that, according to the UN, "if food waste was a country, it would be the #3 global greenhouse emitter."

#4 - Reuse:
Be aware of your purchases and think if you really must buy something new rather than used. Buy a used item on Amazon, go to Goodwill rather than a department store, use Craigslist rather than IKEA. Most of the stuff on Craigslist is from IKEA!

#5 - Recycle correctly:
Take the extra time to put the right item into the right bin. Meaning, if it is recyclable, don't put it into the trash. Discarding a recyclable item into the landfill is the worst case scenario - it will now forever be in the landfill. It can get complicated but do your best!

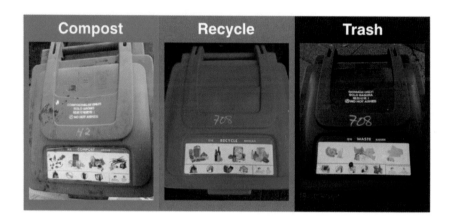

#6 - Adopt composting:
Curbside composting isn't available everywhere but if you've got it, use it. You'll be surprised by how much landfill waste you actually produce if you recycle and compost to the full extent.

#7 - Spread the word:
Get anybody and everybody you know to start recycling. Word of mouth is the best way to get adoption. My close friends and family formed my personal recycling habits.

#5

Is Recycling Good?

There is debate if recycling actually produces an overall negative impact on the environment. The logic lies in the complexity and diversity of current materials. Such a wide array of materials for very specific purposes cannot be efficiently separated so materials are downcycled (used in less valuable products) - reducing recycling effectiveness and, as some argue, creates recycling for recycling sake only.

I saw the complexity of current products firsthand at my first engineering job. My company made plastics from natural gas. We produced over 30 different products with niche applications but all were classified as #5 recyclables. If each of these products were mixed together, the variety in density of the #5 plastic group would be too high to produce material which could be recycled into the products they are intended.

Specifically, plastic in your car bumper is a #5 recyclable but, with current recycling techniques, cannot be remanufactured to create your #5 recyclable tupperware. The density of plastic in a car bumper is too different and would not run on the machinery used to mold tupperware. So, the mix of plastic is downcycled to a product of lower value that can be created with a hodgepodge of #5 recyclable materials.

To extract each sub-type of #5 recyclables would be much, much more expensive than creating a new product from raw materials. With this in mind, the effectiveness of recycling, especially mandated recycling programs, has been debated and even argued to cause more environmental harm than it helps. Daniel Benjamin argues that "since recycling is a manufacturing process, it, like all manufacturing, has its own environmental impact." Personally, I know this is an unpopular opinion but there are various studies to support this view. In college, I wrote a research paper on the effectiveness of reusable water bottles. The paper I based my argument on found that a reusable steel water bottle needs to be used 2000 times in order to equal using a styrofoam cup once (the study claimed the effects on the environment of the styrofoam cup rotting away in the landfill was negligible - which is up for debate and really difficult to calculate).

In my personal view, recycling is like any manufacturing process for a new industry and will improve over time. So, even if recycling is not achieving the goals people believe it accomplishes, we as a society have decided and are investing in better methods to extract more value from our waste. The future isn't certain but I don't doubt people will be able to creatively solve this problem.

References:

#1: *An Introduction to Waste*

https://www.biocycle.net/2004/10/22/increasing-dollar-value-for-compost-products/
https://www.greenbiz.com/article/yes-recycling-still-good-business-if-happens
www3.epa.gov/epawaste/nonhaz/municipal/msw99.htm
www.cleanenergyprojects.com/Landfill-Tipping-Fees-in-USA-2013.html
http://www.bottlebill.org/legislation/usa/california.htm
Blum, B. (1992). Composting and the Roots of Sustainable Agriculture. Agricultural History, 66(2), 171-188. Retrieved from http://www.jstor.org/stable/3743852

#2 *The Evolution of Recycling*

http://cdn2.brooklynmuseum.org/images/opencollection/objects/size4/2000.112.9_bw.jpg
http://cdn.loc.gov/service/pnp/cph/3f00000/3f05000/3f05600/3f05676v.jpg
https://en.wikipedia.org/wiki/Paper_Salvage_1939%E2%80%9350#/media/File:INF3-212_Salvage_I_need_your_waste_paper_(infantry_soldier_figure_calling).jpg
https://upload.wikimedia.org/wikipedia/commons/d/d4/Scrap%5E_Will_Help_Win._Don't_Mix_it_-_NARA_-_533983.jpg
https://en.wikipedia.org/wiki/Mobro_4000
http://www.newsoftheodd.com/article1018.html
http://www.bostonfed.org/economic/nerr/rr2002/q1/waste.htm
https://swana.org/Portals/0/Awards/2012Noms/Recycling_Gold.pdf
https://openlibrary.org/books/OL2037510M/The_Problem_of_waste_disposal
http://www.intergeo.com/en/industry/products_services/waste_management/Organic_Waste_Extruder_Press.html

#3 Case Study: Shoreway Environmental Center – San Mateo County, California, USA

http://www.bulkhandlingsystems.com/equipment/metering-bin/
http://www.bulkhandlingsystems.com/equipment/bag-breaker/
http://www.bulkhandlingsystems.com/equipment/glass-breaker/
http://www.bulkhandlingsystems.com/equipment/nihot-glass-clean-up-system/
http://www.bulkhandlingsystems.com/equipment/newsorter/
http://www.bulkhandlingsystems.com/equipment/polishing-screen/
http://www.bulkhandlingsystems.com/nrt-optical-sorting/
http://www.bulkhandlingsystems.com/equipment/eddy-current-separator/

#4 Seven Things You Can Do

http://www.toxicsaction.org/sites/default/files/tac/information/moving-towards-zero.pdf
https://www3.epa.gov/epawaste/nonhaz/municipal/msw99.htm
http://blogs.scientificamerican.com/plugged-in/un-says-that-if-food-waste-was-a-country-ite28099d-be-the-3-global-greenhouse-gas-emitter/

#5 – Is recycling good?

http://www.bostonfed.org/economic/nerr/rr2002/q1/waste.htm
http://www.perc.org/sites/default/files/ps47.pdf
http://yalefreepress.sites.yale.edu/news/recycling-assumptions-mostly-garbage

Made in the USA
Columbia, SC
06 June 2020

98621079R00020